UNCONSCIOUS ALGORITHMS

MAPPING THE CONFLUENCE OF NEUROSCIENCE AND
UNCONSCIOUS BIAS:ARTIFICIAL INTELLIGENCE AND
THE IMPACT ON BLACK GIRLS AND WOMEN OF GEN Z

DR. MISTY D. FREEMAN

EDITED BY
TAKARA M. CARTER

INCLUSIVE INNOVATOR PUBLISHING

PREFACE

In a world where technology and artificial intelligence (AI) play an increasingly influential role, it is vital that we examine the intersection of race, gender, neuroscience, unconscious bias, and AI. This book aims to shed light on the complex dynamics and implications of this intersection, focusing on the experiences of Gen Z black girls and women. By understanding the historical context, the functioning of the human brain, and the social factors that shape our perceptions, we can gain insights into the biases and inequalities ingrained within our society.

Chapter 1 sets the stage by introducing the main themes and objectives of this book. We explore the fundamental questions surrounding race, gender, neuroscience, unconscious bias, and AI. By establishing a common foundation, we lay the groundwork for the chapters to come.

In Chapter 2, we take a step back to explore the historical context that has shaped the development of science and technology. It becomes apparent that the legacy of racism and sexism still influences our present-day advancements. By understanding this history, we can better navigate the challenges and opportunities that lie ahead.

The intricate relationship between the brain and our perceptions of race and gender is the focus of Chapter 3. We discuss the fascinating field of neuroscience to uncover how our brains process and interpret information related to these identities. By acknowledging the neurological underpinnings, we can gain insight into the origins of unconscious biases.

Chapter 4 takes us beyond the realm of biology to examine the influence of culture and environment on our perceptions. We explore how social factors shape our understanding of race and gender, and how they interact with the neurological processes we discussed earlier. By examining this dynamic interplay, we unravel the complexities of identity and experience.

Intersectionality, a concept often discussed but not fully understood, takes center stage in Chapter 5. We explore the unique challenges faced by Gen Z black girls and women, whose experiences are shaped by the interlocking systems of race, gender, and other social categories. By embracing intersectionality, we can forge a more inclusive path forward.

In Chapter 6, we confront the risks and opportunities posed by AI. While AI has the potential to perpetuate biases and discrimination, it can also be harnessed to address these very issues. We dive into the ethical implications and explore strategies for leveraging AI for positive change.

Chapter 7 emphasizes the importance of representation in AI. We discuss the benefits of diverse voices and perspectives in AI development, and highlight the need for inclusivity to ensure that the technology serves all members of society equitably.

Education and training take center stage in Chapter 8, as we examine the critical role they play in preparing Gen Z black girls and women for a future shaped by AI. By empowering them with the knowledge and skills, we can bridge the digital divide and ensure their active participation in shaping our technological landscape.

Ethics and accountability in AI development are the focal points of Chapter 9. We explore the moral considerations surrounding AI applications, emphasizing the need for robust regulations and frameworks to safeguard against potential harm and ensure fairness and transparency.

In Chapter 10, we embrace the transformative potential of AI to address social inequities. We explore how AI can be leveraged to create a more just and inclusive society, with Gen Z black girls and women playing a pivotal role in driving these transformative changes.

Chapter 11 presents compelling case studies that showcase AI applications successfully promoting diversity and equity. These real-world examples show the tangible impact of inclusive AI development and offer inspiration for future endeavors.

Finally, Chapter 12 brings us to a close, where we summarize the key takeaways from this book and chart a course forward. We offer recommendations for individuals, institutions, and policymakers to collaborate in creating a more fair and inclusive experience for Black girls and women of Gen Z.

INTRODUCTION

The combination of neuroscience, unconscious bias, and artificial intelligence has significant implications for Black girls and women of the Gen Z generation. These three fields of study provide a comprehensive understanding of the unique challenges that Black girls and women face in a world where systemic inequality and discrimination persist.

Neuroscience allows us to examine how the brain processes and responds to different stimuli, ultimately affecting perception and decision-making. For Black girls and women, this can have significant implications. By studying the brain, we can gain insight into how experiences of discrimination and bias can lead to negative outcomes, such as decreased motivation, lowered self-esteem, and limited opportunities.

Unconscious bias is a key factor in perpetuating systemic inequality. By studying how our implicit biases impact our interactions with others, we can break down the barriers that prevent Black girls and women from achieving their full potential.

Finally, artificial intelligence has the potential to either reinforce or challenge existing biases and discrimination. By analyzing the role that technology plays in our lives, we can gain a better understanding of how it can either perpetuate or mitigate inequality.

This book, authored by Dr. Misty D. Freeman, Founder & CEO of Mocha Sprout, explores these intersections and offers insights into how we can use this knowledge to create a more fair society for everyone. Dr. Freeman has worked as a Racial Equity Trainer and Unconscious Bias coach for over 10 years, and her extensive experience in these fields is clear in the raw authenticity of her writing.

Each chapter of the book contains both evidence of occurrences and personal experiences, providing a comprehensive understanding of the issues at hand. Dr. Freeman's goal is to be straightforward and honest about the challenges that Black girls and women face, and to challenge readers to think critically about the implications of these challenges. The content in this book includes not only quantitative studies but also transparent discussions of why the concepts are relevant and important. Overall, this book serves as a call to action for individuals and society to work towards creating a more just and equitable world for all.

UNCONSCIOUS ALGORITHMS

[PART 1]
UNCONSCIOUS ALGORITHMS AND AI BIAS

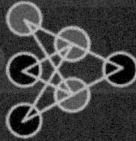

HISTORICAL CONTEXT
THE LEGACY OF RACISM AND SEXISM IN SCIENCE AND TECHNOLOGY

THROUGHOUT HISTORY, SCIENTIFIC AND technological advancements have been shaped by racism and sexism, which has had significant impacts on the development of these fields. These biases have led to the exclusion of certain groups from these fields and have limited the perspectives and ideas that have been used to advance them. For example, the lack of diversity in the early days of computer science led to the development of systems that were biased against women and people of color. Similarly, the exclusion of women from medical research has led to a lack of understanding of how diseases affect women differently than men.

Despite efforts to address these issues, the legacy of discrimination continues to impact the way that science and technology are developed today. For instance, women and people of color continue to be underrepresented in many STEM fields, which limits the diversity of ideas and perspectives that are used to advance these fields. Biases can be unintentionally built into algorithms and other technological systems, perpetuating discrimination and inequality. Therefore, it is important to continue to address these issues and work towards creating more inclusive and equitable scientific and technological advancements.

The Tuskegee Syphilis Study, which took place from 1932 to 1972, is a notorious example of racism and sexism in science and technology. The study, which was conducted by the United States Public Health Service, aimed to observe the natural progression of untreated syphilis in black men. However, the participants were not informed that they had syphilis, and were not given any treatment, even after penicillin became widely available as a cure. The study was highly unethical and caused significant harm to the participants and their families, with many suffering from severe health problems and even death. The implications of this study have been far-reaching and continue to be felt to this day, highlighting the importance of ethical considerations in scientific research and the need to ensure that all participants are treated with dignity and respect.

Another example of the exploitation of people in the medical field is the story of Henrietta Lacks. In 1951, doctors took cancer cells from Lacks without her consent. These cells were then used to create the first immortal cell line, which has been used in countless scientific studies since then. Although Lacks' cells have been a boon to medical research, she and her descendants have never been compensated for their contribution to science. This raises important ethical questions about the use of human tissue in research and the need for informed consent. Lacks' story provides a stark reminder of the historical exploitation of black women in medical research, and the ongoing need for greater equity in healthcare.

These examples illustrate how science and technology can perpetuate systems of racism and sexism. It is important for scientists and researchers to acknowledge the historical and systemic inequalities that have contributed to these issues and work to create more fair and ethical practices in their fields.

The legacy of racism and sexism in science and technology continues to impact the way that these fields are developed today. For example, studies have shown that implicit bias affects the way that scientific studies are designed and conducted. This bias can lead to the underrepresentation of certain groups in research studies and limit the generalizability of study results.

In addition, the lack of diversity in science and technology fields can perpetuate the biases and discrimination of the past. When the people developing technology are not representative of the diverse range of people who use that technology, there is a risk that the technology will not meet the needs of all users. For example, facial recognition technology has been shown to have higher error rates for people with darker skin tones, due to the lack of diversity in the data sets used to train the technology.

In order to address the legacy of racism and sexism in science and technology, it is important to acknowledge and understand this history. This includes recognizing the contributions of historically marginalized groups to these fields and working to provide opportunities for underrepresented groups in science and technology. It is also important to address implicit bias in research and technology development. This can be done by implementing policies and procedures to promote diversity and inclusion in these fields, as well as by providing education and training on implicit bias.

Ultimately, creating a more evenhanded and just society requires a commitment to challenging the legacy of discrimination and bias in science and technology. By working to address these issues, we can create a future where science and technology serve the needs of all people, regardless of race or gender.

UNDERSTANDING THE INTERSECTION OF RACE, GENDER, NEUROSCIENCE, UNCONSCIOUS BIAS, AND ARTIFICIAL INTELLIGENCE

In our rapidly evolving world, advancements in technology and artificial intelligence (AI) have become increasingly intertwined with issues of race and gender. The intersection of race, gender, neuroscience, unconscious bias, and AI presents a complex landscape that demands exploration and understanding. This chapter serves as an introduction to the key themes and objectives of this book, setting the stage for the deeper exploration that follows.

To grasp the profound implications of the intersection of race, gender, neuroscience, unconscious bias, and AI, we must first recognize their interconnectedness. Race and gender are social constructs that shape individuals' experiences, opportunities, and perceptions. Neuroscience reveals how our brains process information related to these identities, shedding light on the origins of unconscious biases. Meanwhile, AI, as a rapidly advancing technology, can both perpetuate and challenge these biases, depending on how it is developed and deployed.

Unconscious biases are deeply ingrained, automatic associations and stereotypes that affect our perceptions and decision-making processes, often without our conscious awareness. They are the product of the social and cultural conditioning we are exposed to from an early age. Studies have shown that even well-intentioned individuals can harbor unconscious biases, impacting their interactions and judgments. For example, research conducted by Greenwald and Krieger (2006) revealed implicit biases towards race and gender, highlighting the pervasive nature of these biases.

Neuroscience plays a crucial role in understanding how our brains process and interpret information related to race and gender, shedding light on the origins of unconscious biases. Neuroimaging studies have showed that certain brain regions involved in perception, emotion, and social cognition are activated when processing racial and gender information (Hart et al., 2000; Freeman & Ambady, 2011). These findings suggest that our brains have inherent mechanisms that influence our perceptions and judgments related to race and gender.

Artificial intelligence, with its ability to analyze vast amounts of data and make autonomous decisions, has the potential to amplify or challenge societal biases. However, AI systems are only as unbiased as the data they are trained on and the algorithms used. If AI systems are trained on biased data or developed without diverse perspectives, they can perpetuate and even amplify existing biases. For instance, facial recognition algorithms have been shown to have higher error rates for women and people with darker skin tones (Buolamwini & Gebru, 2018). This highlights the urgent need to address bias in AI development and deployment.

The primary objective of this book is to deepen our understanding of the intersection of race, gender, neuroscience, unconscious bias, and AI. By exploring these topics, we aim to uncover the ways in which biases emerge, persist, and can be mitigated in AI development. We seek to highlight the experiences and perspectives of Gen Z black girls and women, who are often marginalized in both technological spaces and broader society.

By examining historical context, understanding the workings of the brain, and considering the role of culture and environment, we can gain insights into the complex dynamics at play. We will also explore the concept of intersectionality and its impact on the experiences of Gen Z black girls and women. We will examine the risks and opportunities of AI, emphasizing the importance of diversity and representation in its development.

The intersection of race, gender, neuroscience, unconscious bias, and AI presents a multifaceted landscape that demands exploration and understanding. By delving into these themes and examining their interconnections, we can better navigate the complexities of bias, inequality, and the potential for transformative change.

UNRAVELING THE THREADS OF DISCRIMINATION

RACISM AND SEXISM'S ENDURING IMPACT ON SCIENCE AND TECHNOLOGY

To FULLY UNDERSTAND THE CHALLENGES FACED AT the intersection of race, gender, neuroscience, unconscious bias, and artificial intelligence, it is essential to examine the historical context that has shaped the development of science and technology. This chapter explores the deep-rooted legacy of racism and sexism, shedding light on how historical factors continue to influence contemporary advancements. By delving into this history, we can gain insights into the systemic issues that persist and their impact on marginalized communities.

The fields of science and technology have not existed in a vacuum, untouched by the societal prejudices of their time. Instead, they have been deeply entwined with the prevailing ideologies and power structures. From the early days of scientific exploration to the development of modern technologies, racism and sexism have influenced research, discoveries, and the allocation of resources.

One significant aspect of this legacy is the exclusion of marginalized communities from scientific spaces. Historically, access to education, funding, and professional opportunities has been limited for individuals from racial and gender minorities. This exclusion has had long-lasting effects on representation and the ability of marginalized groups to contribute to scientific advancements.

Throughout history, scientific racism and gender essentialism have been used to justify discriminatory practices and reinforce societal hierarchies. In anthropology, for example, racial classifications and theories of racial superiority or inferiority were used to promote discriminatory policies and support colonial expansion. Similarly, gender essentialism perpetuated the belief that women were inherently less capable in certain domains, leading to the marginalization of women in scientific and technological fields.

The history of science and technology is marred by ethical violations and unethical experiments, particularly towards marginalized communities. Examples such as the Tuskegee Syphilis Study, which targeted African American men and the forced sterilization of women of color, highlight the systemic abuse and exploitation endured by these communities. These egregious acts underscore the need for ethical frameworks and safeguards to protect vulnerable populations.

Amidst the deeply entrenched biases and discriminatory practices, there have been individuals and movements that have fought against the status quo. Pioneering figures like Marie Curie, who challenged gender norms in science, and the Civil Rights Movement, which fought for racial equality, have paved the way for progress. However, it is crucial to recognize that progress has been uneven, and the effects of historical biases persist to this day.

The historical legacy of racism and sexism in science and technology continues to shape contemporary developments. The underrepresentation of racial and gender minorities in STEM fields, biased algorithms, and the perpetuation of harmful stereotypes are just a few examples of how historical biases persist. To address these issues, it is essential to acknowledge and confront the historical context that has shaped the current landscape.

Chapter 2 provides a glimpse into the historical context of racism and sexism in science and technology. By recognizing the systemic issues and biases that have shaped the field, we can understand the challenges faced at the intersection of race, gender, neuroscience, unconscious bias, and artificial intelligence.

This historical analysis serves as a foundation for further exploration in subsequent chapters, where we will examine the current state of affairs and seek pathways towards a more nondiscriminatory and inclusive future.

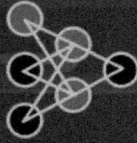

UNDERSTANDING THE BRAIN

HOW NEUROSCIENCE SHAPES OUR PERCEPTIONS OF RACE AND GENDER

TO COMPREHEND THE INTRICATE DYNAMICS AT THE intersection of race, gender, neuroscience, unconscious bias, and artificial intelligence, it is crucial to explore how the brain processes information related to these aspects of identity. Chapter 3 uncovers the fascinating realm of neuroscience, uncovering how our neural processes influence our perceptions, judgments, and biases regarding race and gender. By unraveling these cognitive mechanisms, we can shed light on the origins of unconscious bias and its profound implications.

Neuroscience research has revealed that our brains engage in distinct processes when perceiving and processing information related to race and gender. Through neuroimaging techniques such as functional magnetic resonance imaging (fMRI), researchers have identified specific brain regions involved in these cognitive processes. For instance, the fusiform face area (FFA) plays a role in facial recognition, while the amygdala, involved in emotional processing, can be activated when processing racial and gender cues (Hart et al., 2000; Freeman & Ambady, 2011).

As our brains develop, they undergo a process of neuroplasticity, where experiences and environmental factors shape neural connections and cognitive processes. Research suggests that early exposure to racial and gender cues, societal messages, and cultural norms can impact the development of neural networks associated with race and gender perception (Shutts et al., 2016). These experiences can contribute to the formation of unconscious biases and influence subsequent judgments and behavior.

To measure the presence and strength of unconscious biases, scientists have developed implicit association tests (IATs). IATs utilize response times and associations between different concepts to gauge the strength of implicit biases individuals may hold. Studies using IATs have consistently showed implicit biases related to race and gender, even among individuals who consciously hold egalitarian beliefs (Greenwald & Krieger, 2006). These findings underscore the subtle nature of unconscious biases and their potential influence on our thoughts and actions.

Stereotypes, deeply ingrained beliefs and assumptions about social groups, can significantly impact our perceptions and judgments. Neuroscience research has shown that stereotypes can influence neural processes related to the perception of race and gender. For instance, exposure to stereotypes can influence the activation of brain regions associated with emotional processing and attention, shaping subsequent cognitive processes and behavior (Cvencek et al., 2011; Amodio, 2014).

While neuroscience reveals the neural underpinnings of biases, it also provides hope for bias mitigation. The concept of neuroplasticity highlights the brain's ability to reorganize and form new connections throughout our lives. By understanding this capacity for change, interventions can be developed to promote bias reduction and foster more inclusive attitudes and behaviors. Research has showed the effectiveness of interventions such as perspective-taking exercises, empathy training, and exposure to counter-stereotypical information in reducing unconscious biases (Dasgupta & Greenwald, 2001; Lai et al., 2014).

Chapter 3 has explored the fascinating realm of neuroscience, shedding light on how our brains process and perceive race and gender. By unraveling the neural mechanisms at play, we can begin to understand the origins of unconscious biases and their impact on our perceptions and judgments. This knowledge sets the stage for further exploration in subsequent chapters, where we will look deeper into the societal and cultural factors that interact with neuroscience to shape our understanding of race, gender, and bias. Through this exploration, we aim to foster a greater understanding of the complexities at the intersection of these fields and seek pathways towards promoting equity and inclusivity in our society.

THE ROLE OF CULTURE AND ENVIRONMENT
HOW SOCIAL FACTORS AFFECT PERCEPTION

CHAPTER 4 RECOUNTS PROFOUND INFLUENCE OF culture and environment on our perceptions of race and gender. By examining the complex interplay between societal factors and neurological processes, we can gain a deeper understanding of how cultural contexts shape our biases and attitudes. Recognizing the significance of these social factors is essential in unraveling the complexities of intersectionality and addressing the unique challenges faced by marginalized communities.

Cultural norms and stereotypes play a significant role in shaping our perceptions and judgments. Society's beliefs and values surrounding race and gender influence the formation of stereotypes, which in turn impact how we perceive and interact with others. For example, societal narratives that associate certain racial or ethnic groups with specific characteristics or roles can contribute to the formation of biased beliefs and attitudes (Devine, 1989).

The process of socialization plays a crucial role in the development of implicit biases. From a young age, individuals are exposed to societal messages, media representations, and interpersonal experiences that shape their understanding of race and gender. These influences can lead to the internalization of biased attitudes and stereotypes, even without conscious awareness (Blair, 2002). Consequently, implicit biases can become deeply ingrained and perpetuate inequalities in various domains.

Stereotype threat refers to the phenomenon where individuals, aware of negative stereotypes associated with their social group, experience anxiety and reduced performance in domains related to those stereotypes (Steele & Aronson, 1995). For example, the stereotype that women are less proficient in math can create a psychological burden that hinders their performance in mathematical tasks. This shows the detrimental impact of societal expectations on individuals' abilities and opportunities.

Intersectionality recognizes individuals embody multiple social identities that interact and intersect to shape their experiences. The experiences of Gen Z black girls and women, for instance, are influenced not only by their gender but also by their race, socioeconomic status, and other factors. Understanding the unique challenges faced by individuals at the intersections of multiple identities is essential in addressing the complexities of systemic inequality (Crenshaw, 1989).

Cultural competence refers to the ability to understand, appreciate, and effectively interact with individuals from diverse cultural backgrounds. Developing cultural competence is crucial in dismantling biases and fostering inclusive environments. Empathy plays a key role in this process, as it allows us to step into the shoes of others and gain a deeper understanding of their experiences (Galinsky et al., 2013). Cultivating empathy and cultural competence can contribute to more fair interactions and promote positive social change.

Chapter 4 has highlighted the significant influence of culture and environment on our perceptions of race and gender. By recognizing the role of cultural norms, stereotypes, socialization, and intersectionality, we can better understand how these factors shape our biases and attitudes. Acknowledging the power of cultural competence and empathy, we can cultivate inclusive environments that honor diverse experiences and promote equity. In the subsequent chapters, we will explore the impact of artificial intelligence and unconscious bias in perpetuating or challenging these societal dynamics, and the importance of representation and education in fostering positive change.

INTERSECTIONALITY

UNDERSTANDING THE COMPLEXITY OF IDENTITY AND
EXPERIENCE

CHAPTER 5 EXPLORES THE CONCEPT OF intersectionality and its profound impact on the experiences of Gen Z black girls and women. Intersectionality recognizes individuals exist at the intersection of multiple social identities, such as race, gender, class, and sexuality, which interact and shape their lived experiences. Understanding the complexities of intersectionality is essential in addressing the unique challenges faced by marginalized communities and promoting a more inclusive and nondiscriminatory society.

Intersectionality emphasizes that identity is multifaceted and cannot be reduced to a single dimension. Gen Z black girls and women navigate a complex web of identities and face unique experiences that arise from the intersection of their race, gender, and other social identities. For example, their experiences may be shaped by racism, sexism, and the interplay of these forms of discrimination. Recognizing these intersecting identities is crucial in understanding the complexities of their lived realities.

Intersectionality sheds light on power dynamics and systems of oppression that operate within society. It acknowledges that individuals with intersecting marginalized identities may face compounded forms of discrimination and disadvantage. For example, the experiences of a black woman may differ from those of a white woman or a black man due to the intersection of race and gender. Understanding these power dynamics is vital in addressing systemic inequities and promoting social justice.

Gen Z black girls and women encounter a range of challenges resulting from the intersection of their identities. They may face barriers in educational and professional settings, experience stereotype threat, and contend with systemic racism and sexism. Despite these challenges, many show remarkable resilience and strength. Recognizing their resilience and amplifying their voices is essential in challenging societal norms and empowering marginalized communities.

Representation matters, as it shapes our perceptions and understanding of diverse identities. Positive and authentic representation of Gen Z black girls and women in media, education, and leadership positions is crucial in challenging stereotypes and promoting inclusivity. By increasing their visibility, we can create spaces where their experiences and perspectives are valued and respected.

Amplifying the voices of Gen Z black girls and women and fostering allyship are crucial steps toward dismantling systemic inequities. Listening to and centering their experiences, concerns, and aspirations can drive meaningful change. Allies play a vital role in challenging biases, advocating for equity, and creating inclusive spaces that celebrate diversity. Building authentic relationships and engaging in intersectional dialogue can foster understanding and solidarity.

Chapter 5 has explored the concept of intersectionality and its significance in understanding the experiences of Gen Z black girls and women. By recognizing the complexities of intersecting identities, power dynamics, and resilience, we gain insight into the challenges they face and the strength they embody. Through representation, amplification of voices, and allyship, we can create a more inclusive society that values and uplifts individuals at the intersections of multiple identities. In the subsequent chapters, we will discuss the intersection of artificial intelligence, bias, and representation, highlighting the potential for positive change and the importance of diversity in shaping AI technologies.

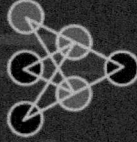

AI AND BIAS

UNDERSTANDING THE RISKS AND OPPORTUNITIES

CHAPTER 6 EXPLORES THE INTRICATE relationship between artificial intelligence (AI) and bias. While AI holds tremendous potential to revolutionize various aspects of our lives, it also presents risks and challenges, particularly in relation to perpetuating biases and discrimination. By delving into the intersection of AI and bias, we can gain a deeper understanding of the complexities involved and explore strategies for mitigating bias and harnessing the transformative power of AI for positive change.

AI systems are not inherently biased, but they can reflect and amplify existing biases present in the data used to train them. Biased data, whether it stems from historical societal biases or data collection processes, can lead to biased outcomes. For example, if a facial recognition system is primarily trained on data that is predominantly male and lighter-skinned, it may exhibit higher error rates when identifying faces of women or individuals with darker skin tones (Buolamwini & Gebru, 2018).

AI bias has a disproportionate impact on marginalized communities, exacerbating existing inequalities. The reliance on biased algorithms can perpetuate discriminatory practices in various domains, including criminal justice, employment, and lending. For instance, predictive policing algorithms have been found to disproportionately target minority communities, perpetuating the over-policing and surveillance of these populations (O'Neil, 2016).

Unintended consequences can arise when biased AI systems are deployed without proper consideration of their social implications. These consequences can range from reinforcing stereotypes to creating exclusionary practices. For example, biased recommendation algorithms in online platforms can reinforce echo chambers and limit individuals' exposure to diverse perspectives (Gillespie, 2018).

Addressing bias in AI requires a multi-faceted approach involving diverse stakeholders, including researchers, developers, policymakers, and communities affected by biased algorithms. Strategies for mitigating bias include:

1. Data Collection and Preprocessing:
   Ensuring diverse and representative
   datasets and implementing rigorous

preprocessing techniques can help reduce biases in AI systems.

2. Algorithmic Transparency and Explainability: Making AI algorithms transparent and explainable can enhance accountability and facilitate the identification and mitigation of bias.

3. Ethical Guidelines and Standards: Developing and adhering to ethical guidelines and standards can ensure responsible AI development and deployment, minimizing the risk of biased outcomes.

4. Diversity and Inclusion in AI Development: Promoting diversity and inclusion in AI research and development teams can lead to more robust and unbiased algorithms that cater to a broader range of perspectives.

5. Continuous Monitoring and Evaluation: Regular monitoring and evaluation of AI systems can help identify and rectify biases that may emerge over time.

6. Collaboration and Community Engagement: Engaging with affected communities and incorporating their perspectives throughout the AI development process can foster more inclusive and equitable systems.

Chapter 6 has examined the intersection of AI and bias, highlighting the risks and challenges associated with biased AI systems. By recognizing the impact on marginalized communities and understanding the unintended consequences, we can develop strategies to address bias in AI. Through data collection practices, algorithmic transparency, ethical guidelines, diversity and inclusion, continuous monitoring, and community engagement, we can work towards creating AI systems that are more equitable, accountable, and beneficial for all. In the subsequent chapters, we will look deeper into the importance of representation, diversity, and inclusion in AI and explore the potential for AI to address social inequities.

THE IMPORTANCE OF REPRESENTATION

DIVERSITY AND INCLUSION IN AI

CHAPTER 7 EMPHASIZES THE CRITICAL importance of representation, diversity, and inclusion in the field of artificial intelligence (AI). The lack of diversity within AI development teams and the under-representation of marginalized communities have significant implications for the fairness and inclusivity of AI systems. By exploring the benefits of diverse representation and the challenges that hinder inclusiv-ity, we can recognize the imperative for creating AI systems that reflect the diversity of the world they are designed to serve.

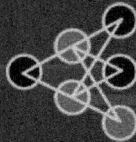

Representation matters in AI. Diverse representa-tion within AI development teams brings a variety of perspectives and experiences to the table, enabling the creation of more inclusive and equitable systems. When individuals from different backgrounds, cultures, and identities are involved in the design and

decision-making processes, AI technologies are more likely to address the needs and interests of a broader range of users (Gebru et al., 2020).

Diverse teams can play a pivotal role in mitigating bias and discrimination within AI systems. By bringing diverse perspectives to the forefront, biases embedded in data, algorithms, and models can be identified and rectified more effectively. Without diverse representation, AI systems may perpetuate existing biases and contribute to further discrimination against marginalized communities (Benjamin, 2019).

Data bias is a significant concern in AI systems. Biases present in training data can lead to biased outcomes and discriminatory decisions. Diverse representation is crucial in curating and validating training datasets to ensure they are inclusive and representative of different identities and experiences. By actively involving individuals from underrepresented groups, AI developers can reduce the risk of perpetuating biases in their systems (Hao, 2020).

Inclusive AI requires a user-centered design approach that considers the diverse needs, preferences, and experiences of all users. By involving individuals from marginalized communities in the design process, AI systems can be tailored to address their specific concerns and challenges. User feedback and iterative design cycles can lead to more accessible, inclusive, and user-friendly AI technologies (DiSalvo, 2012).

Diversity and inclusion in AI have broader societal

implications beyond technology itself. Increasing representation in AI can serve as a catalyst for breaking systemic barriers and fostering equity in education, employment, and opportunities. By creating pathways for underrepresented individuals to enter the field of AI and providing support networks, we can work towards dismantling structural inequalities and promoting social mobility (Crawford et al., 2019).

Chapter 7 has emphasized the critical importance of representation, diversity, and inclusion in AI. By recognizing the power of diverse perspectives, the need to address bias, and the benefits of user-centered design, we can create AI systems that are more inclusive, equitable, and accountable. The involvement of underrepresented communities in AI development is essential for mitigating bias, fostering innovation, and promoting social justice. In the subsequent chapters, we will explore the role of education and training in equipping Gen Z black girls and women for the AI-driven future, and the ethical considerations and regulatory frameworks necessary to ensure responsible AI development.

EDUCATION AND TRAINING

EQUIPPING GEN Z BLACK GIRLS AND WOMEN FOR THE FUTURE

CHAPTER 8 EXPOSES THE CRUCIAL ROLE OF education and training in preparing Gen Z black girls and women for the challenges and opportunities presented by an AI-driven future. Recognizing the transformative potential of AI technologies, it becomes imperative to provide restitution access to quality education and empower individuals with the skills to actively participate and shape the AI landscape. By examining the barriers and advocating for inclusive educational practices, we can foster the development of a diverse and empowered generation ready to navigate the AI-powered world.

Gen Z black girls and women often face barriers to accessing quality education and opportunities in the AI field. These barriers can include limited resources, lack of representation, and systemic biases that perpetuate educational inequalities. It is essential to address these barriers and create inclusive learning environments that promote equal access and representation for all learners (Lopez, 2019).

STEM (Science, Technology, Engineering, and Mathematics) education plays a vital role in preparing individuals for AI-driven careers. However, there is a need to actively promote and support STEM education among Gen Z black girls and women. This can be achieved through targeted initiatives, mentorship programs, and partnerships with educational institutions and organizations that encourage interest and participation in STEM fields (Archer et al., 2019).

As AI technologies continue to advance, it is crucial to educate Gen Z black girls and women about the ethical and social implications of AI. This includes understanding issues related to bias, privacy, algorithmic transparency, and the impact of AI on various aspects of society. By fostering critical thinking skills and ethical awareness, individuals can actively engage in shaping the development and deployment of AI in ways that align with social justice and equity (Jobin et al., 2019).

To thrive in an AI-driven future, Gen Z black girls and women need to develop AI-related skills. This includes computational thinking, data literacy, programming, and understanding AI principles. Integrating AI education into the curriculum, providing access to coding resources, and offering mentorship and internship opportunities can empower learners to become creators and innovators in the AI landscape (Wing, 2006).

Education should also focus on raising awareness about bias in AI systems and fostering algorithmic accountability. By understanding how bias can manifest in AI technologies and its impact on marginalized communities, learners can actively contribute to mitigating bias and ensuring the development of fair and unbiased AI systems. Promoting transparency and accountability in algorithmic decision-making processes can foster trust and encourage responsible AI practices (Gillespie, 2020).

Inclusive AI education requires an intersectional approach that addresses the unique experiences and challenges faced by Gen Z black girls and women. This includes incorporating diverse perspectives, histories, and cultural contexts into AI curricula, as well as ensuring that learning materials and resources are inclusive and accessible to learners from diverse backgrounds. By creating a supportive and inclusive learning environment, we can empower Gen Z black girls and women to fully participate in the AI-driven future (Jackson, 2020).

Chapter 8 has highlighted the importance of education and training in equipping Gen Z black girls and women for the AI-driven future. By addressing barriers to access and representation, promoting STEM education, fostering ethical awareness, equipping learners with AI skills, and promoting inclusive

AI education, we can empower this generation to become active participants and leaders in shaping the future of AI. In the subsequent chapter, we will look into the ethical considerations and the role of regulation in AI development to ensure responsible and equitable AI practices.

The history of science and technology is fraught with examples of racism and sexism. From the exclusion of women and people of color from scientific and technological advancements, to the use of science and technology to perpetuate oppression and discrimination, the legacy of these biases is still present today.

One example of this legacy is the history of eugenics, which sought to "improve" the genetic quality of the human population through selective breeding and sterilization. The eugenics movement in the United States gained significant traction in the early 20th century, with many prominent scientists and politicians advocating for its implementation. However, eugenics was often used to justify racist and ableist policies, including forced sterilization of people of color, disabled individuals, and those deemed "unfit" to reproduce.

Similarly, the history of psychology and psychiatry is filled with examples of racial and gender bias. For instance, the concept of "hysteria" was used to pathologize women's emotions and behavior, and was often treated with highly invasive and unnecessary medical procedures. In addition, the field of psychology has a history of using white, male participants in studies, which has led to a lack of understanding of how race and gender impact psychological processes.

The field of technology also has a long history of excluding women and people of color. For example, the development of the personal computer was largely dominated by men, and women who did contribute to its development were often marginalized and excluded from recognition. In addition, many technologies have been designed with implicit biases that can perpetuate discrimination, such as facial recognition software that is less accurate for people with darker skin tones.

These examples illustrate how the legacy of racism and sexism in science and technology has shaped our understanding of the world and perpetuated inequality. It is crucial to acknowledge and address this legacy in order to create a more inclusive and just future.

Dr. Dorothy E. Roberts, a scholar of race, gender, and the law, has argued that addressing these biases requires a shift in our approach to science and technology. In her book, "Fatal Invention: How Science, Politics, and Big Business Re-create Race in the Twenty-First Century," Roberts argues we must move away from a "color-blind" approach to science and technology, and instead recognize and address the ways in which racism and sexism are built into these fields. By doing so, we can create technologies and policies that are inclusive.

The history of science and technology is deeply

intertwined with racism and sexism, and this legacy continues to impact our understanding of the world today. It is crucial to acknowledge and address this legacy in order to create a more fair and just future. By recognizing and addressing the ways in which these biases are built into science and technology, we can create more inclusive and restitutionary technologies and policies that benefit everyone.

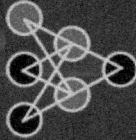

ETHICS AND ACCOUNTABILITY

THE ROLE OF REGULATION IN AI DEVELOPMENT

CHAPTER 9 EXPOUNDS ON THE CRUCIAL ROLE OF ethics and accountability in the development and deployment of artificial intelligence (AI). As AI technologies continue to advance and permeate various aspects of society, it becomes imperative to establish robust regulatory frameworks that ensure responsible and ethical AI practices. By examining the ethical considerations associated with AI, exploring the challenges of accountability, and advocating for regulatory measures, we can work towards harnessing the potential of AI while safeguarding against potential risks and harms.

AI raises a host of ethical considerations that need to be carefully addressed. One significant concern is the potential for bias and discrimination in AI systems. Bias can be unintentionally embedded in algorithms, training data, or decision-making processes, leading to unjust or discriminatory outcomes. It is essential to develop AI systems that are fair, transparent, and accountable, mitigating bias and ensuring equitable treatment for all individuals (Boddington, 2017).

Another ethical concern is privacy and data protection. AI technologies often rely on vast amounts of data, raising concerns about how personal information is collected, stored, and used. It is crucial to establish robust data protection regulations and mechanisms for informed consent to protect individuals' privacy rights and prevent misuse or unauthorized access to sensitive information (Jobin et al., 2019).

The lack of transparency and explainability in AI systems poses significant ethical challenges. As AI algorithms become more complex, it becomes difficult to understand how they arrive at certain decisions or recommendations. This opacity raises concerns regarding accountability and the potential for unjust outcomes. Developing AI systems that are transparent, explainable, and accountable is essential to build trust and ensure that decisions made by AI are understandable and justifiable (Wachter et al., 2017).

Determining accountability in AI systems can be complex due to the distributed nature of decision-making and the involvement of various stakeholders. It is essential to establish clear lines of responsibility, ensuring that individuals and organizations involved in AI development and deployment are held accountable for the ethical implications of their systems. This includes defining legal and ethical frameworks that outline the rights and responsibilities of developers, users, and the AI systems themselves (Floridi et al., 2018).

Developing robust regulatory frameworks is crucial for ensuring responsible AI practices. Governments and international bodies increasingly recognize the need for regulations that address the ethical, legal, and social implications of AI. These regulations should strike a balance between fostering innovation and protecting against potential harms. They should encompass issues such as bias mitigation, data protection, privacy, transparency, accountability, and the ethical use of AI in various domains (European Commission, 2021).

Given the global nature of AI and its potential impact, international collaboration is vital in developing cohesive regulatory frameworks. Collaboration can facilitate the sharing of best practices, aligning ethical standards, and addressing the challenges posed by cross-border AI applications. By working together, nations can establish a unified approach to AI governance that upholds ethical principles, protects human rights, and ensures that AI benefits all of humanity (OECD, 2019).

Chapter 9 has highlighted the ethical considerations and the need for accountability in AI development. By addressing bias, ensuring transparency and explainability, and establishing regulatory frameworks, we can foster responsible AI practices. It is essential to strike a balance between innovation and ethics, considering the potential risks and harms associated with AI technologies. By working collaboratively and globally, we can build a future where AI serves the best interests of society while upholding fundamental values of fairness, justice, and human rights. In the subsequent chapter, we will explore the potential of AI to address social inequities.

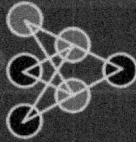

TRANSFORMATIVE TECHNOLOGY

THE POTENTIAL FOR AI TO ADDRESS SOCIAL INEQUITIES

CHAPTER 10 EXPLORES THE TRANSFORMATIVE potential of artificial intelligence (AI) to address social inequities. While AI has been implicated in perpetuating biases and discrimination, it also offers opportunities to tackle systemic injustices and create positive change. By examining the applications of AI in various domains, we can explore how it can be harnessed to promote equity, inclusivity, and social progress.

Access to quality education is a fundamental right, yet it remains elusive for many marginalized communities. AI can play a pivotal role in addressing educational inequities by providing personalized learning experiences, adaptive assessments, and targeted interventions. By leveraging AI technologies, educators can tailor instruction to individual students' needs, bridge learning gaps, and ensure equal opportunities for all learners, regardless of their backgrounds (Freeman, 2022).

Health disparities persist, with certain communities experiencing unequal access to healthcare resources and services. AI can enhance healthcare equity by facilitating early detection of diseases, improving diagnostic accuracy, and supporting precision medicine. By analyzing vast amounts of patient data, AI algorithms can identify patterns, predict health risks, and enable proactive interventions, reducing disparities in healthcare outcomes (Topol, 2019).

The criminal justice system has long been plagued by biases and inequalities. AI applications, such as predictive policing and risk assessment algorithms, have the potential to exacerbate these issues. However, when designed and implemented with care, AI can also contribute to criminal justice reform. For instance, AI can help identify and rectify biased sentencing practices, facilitate fairer bail decisions, and support rehabilitation efforts by providing personalized intervention strategies (Angwin et al., 2018).

Economic disparities are a significant challenge worldwide, and AI can be a powerful tool for promoting economic empowerment. AI-based platforms and marketplaces can create opportunities for underrepresented entrepreneurs and small businesses by connecting them with customers and resources. AI-driven automation can enhance productivity and efficiency in industries, leading to job creation and economic growth. By leveraging AI for inclusive economic development, we can strive towards reducing income inequality and fostering economic opportunities for all (World Economic Forum, 2020).

Addressing environmental challenges, such as climate change and resource depletion, is crucial for a sustainable future. AI can contribute to environmental sustainability by enabling data-driven decision-making, optimizing energy consumption, and supporting conservation efforts. For example, AI algorithms can analyze environmental data to identify patterns and trends, facilitate precision agriculture, and enhance renewable energy systems. By harnessing the power of AI in environmental initiatives, we can work towards a greener and more sustainable planet (Alam et al., 2021).

While the potential of AI to address social inequities is promising, it is crucial to consider the ethical implications and potential risks. Bias, fairness, transparency, and accountability must be at the forefront of AI development and deployment. Careful attention should be paid to data collection and training processes to avoid perpetuating existing biases. Regulatory frameworks and guidelines should ensure responsible AI practices that prioritize social equity and safeguard against unintended harm (Jobin et al., 2019).

Chapter 10 has explored the transformative potential of AI in addressing social inequities across various domains. By leveraging AI for access to education, healthcare equity, social and criminal justice reform, economic empowerment, and environmental sustainability, we can make significant strides towards a more equitable society. However, it is essential to approach AI deployment with a critical eye, considering the ethical implications and potential risks. By combining technological advancements with a commitment to social equity, we can harness the power of AI to create a future that is fair, inclusive, and beneficial for all. In the final chapter, we will present case studies of AI applications that have successfully promoted diversity and equity, illustrating the tangible impact of responsible AI implementation.

[PART 2]
CASE STUDIES

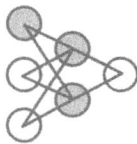

CASE STUDY 1

AI-POWERED HIRING PRACTICES

IN THIS CASE STUDY, WE EXPLORE THE USE OF AI IN hiring practices to mitigate bias and promote diversity in the workplace. Traditional hiring processes often suffer from unconscious biases that can lead to the underrepresentation of certain groups. AI algorithms can be used to screen job applicants objectively, focusing on skills and qualifications rather than demographic factors. By removing biased decision-making and promoting meritocracy, AI-powered hiring practices have the potential to enhance diversity, increase representation, and create more inclusive work environments (Dastin, 2018).

The pursuit of diversity and inclusivity in the workplace has become a paramount goal for organizations across various industries. Traditional hiring practices, while well-intentioned, often fall prey to unconscious biases that unintentionally perpetuate underrepresentation of certain groups. As the world becomes increasingly aware of these disparities, the application of artificial intelligence (AI) in hiring processes offers a promising solution. In this case study, we explore how AI is revolutionizing hiring practices by mitigating bias and promoting diversity in the workplace. By introducing objectivity into the screening process, AI algorithms enable a focus on skills and qualifications, thereby fostering meritocracy and creating more inclusive work environments.

The Challenge of Bias in Traditional Hiring

Traditional hiring practices are vulnerable to unconscious biases, which are deeply ingrained attitudes or stereotypes that influence decision-making subconsciously. Such biases can be based on various factors, including gender, race, ethnicity, age, and socioeconomic background. Despite sincere efforts to eliminate discrimination, these biases can still seep into the hiring process, leading to unintentional underrepresentation and a lack of diversity within organizations.

- Gender Bias: Studies have shown that gender bias often affects evaluations of candidates, leading to the underrepresentation of women in certain industries and leadership roles.

- Racial and Ethnic Bias: Racial and ethnic bias can result in the exclusion of qualified candidates from diverse backgrounds, perpetuating homogeneous workforces.
- Age Bias: Older and younger candidates may face discrimination based on stereotypes associated with their respective age groups.
- Socioeconomic Bias: Socioeconomic bias can inadvertently impact candidates from different economic backgrounds, leading to uneven opportunities for advancement.

AI-Powered Hiring: A Step Towards Objectivity

AI algorithms have the potential to disrupt traditional hiring practices by removing subjective judgments and introducing a data-driven, unbiased approach to candidate evaluation. Through machine learning, AI algorithms can analyze vast amounts of historical hiring data, learning from past decisions to make more objective assessments of future applicants. By focusing solely on skills, qualifications, and experiences, AI can mitigate bias and foster a more inclusive hiring process.

- Skills-Based Screening: AI algorithms can be trained to recognize and prioritize the key skills and qualifications required for a particular job. By concentrating on these objective criteria, AI ensures that candidates are assessed solely on their abilities, rather than on extraneous factors.
- Blind Recruitment: To further eliminate bias, some AI-powered platforms adopt a "blind recruitment" approach. This involves anonymizing candidate information, such as names, photos, and demographic details, during the initial screening stages, ensuring that evaluators are not influenced by unconscious biases.
- Standardized Evaluations: AI algorithms consistently apply the same evaluation criteria to all candidates, resulting in a standardized and fair screening process. This objectivity ensures that candidates are assessed on an equal footing, regardless of their background.

Benefits of AI in Hiring for Diversity

- Mitigating Bias and Increasing Diversity: By removing human subjectivity, AI helps organizations overcome unconscious biases that have hindered diversity efforts. This leads to a more inclusive and diverse workforce, representing a broader range of perspectives and ideas.

- Enhanced Meritocracy: AI-driven hiring practices prioritize skills and qualifications, promoting a merit-based approach that rewards talent and potential. This meritocracy encourages candidates to focus on their abilities, fostering a more competitive and dynamic job market.

- Faster and More Efficient Hiring: AI-powered systems can streamline the hiring process, analyzing large volumes of applications quickly and accurately. This efficiency reduces the time-to-hire, allowing organizations to secure top talent before competitors.

- Improved Retention Rates: A diverse and inclusive workforce fosters a sense of belonging and acceptance, contributing to higher employee satisfaction and reduced turnover rates.

- Meeting Legal and Ethical Obligations: Employing AI to screen candidates can demonstrate an organization's commitment to fair hiring practices, helping them meet legal and ethical requirements related to diversity and equal opportunities.

Challenges and Ethical Considerations

While AI-powered hiring practices offer significant benefits, there are also challenges and ethical considerations that must be addressed.

- Data Bias: AI algorithms learn from historical data, which may contain biases from past hiring decisions. If not carefully addressed, these biases can be perpetuated in the AI's decision-making process.
- Lack of Transparency: Some AI algorithms, particularly complex deep learning models, can be difficult to interpret. This lack of transparency raises concerns about accountability and the potential for unintended bias.
- Technical Limitations: AI algorithms may struggle with assessing certain soft skills or emotional intelligence, which are vital for certain roles but challenging to measure objectively.
- Fairness and Privacy: Ensuring that AI systems are fair to all candidates while respecting their privacy is crucial.

Handling candidate data responsibly and maintaining data privacy are critical aspects of ethical AI implementation.

AI-powered hiring practices have emerged as a transformative solution to address bias in traditional hiring processes and promote diversity in the workplace. By introducing objectivity, standardization, and skill-based screening, AI algorithms can help organizations recruit the most qualified candidates while fostering a more inclusive and diverse workforce. Nevertheless, developers and organizations must remain diligent in addressing potential biases within AI systems and adhering to ethical guidelines to ensure fair and responsible use. As AI technology continues to evolve, its integration into hiring practices will play a pivotal role in advancing diversity and creating more inclusive workplaces that harness the full potential of human talent.

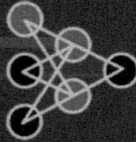

12

CASE STUDY 2

AI FOR LANGUAGE TRANSLATION AND ACCESSIBILITY

LANGUAGE BARRIERS CAN BE FORMIDABLE obstacles, limiting effective communication and access to information for individuals who speak different languages. In an increasingly globalized world, the ability to overcome these barriers is essential for fostering understanding and inclusivity. Artificial intelligence (AI) has emerged as a powerful tool to address language barriers and enhance accessibility for diverse populations. This case study explores the transformative impact of AI in language translation and accessibility, showcasing how AI-powered tools facilitate effective communication across language boundaries and promote inclusivity by aiding individuals with disabilities. (Leidner et al., 2020).

AI-Powered Language Translation

AI-powered language translation tools have revolutionized the way people communicate and access information across different languages. These tools encompass a wide range of applications, from mobile apps to real-time translation services, all working towards bridging language gaps.

- Language Translation Apps: AI-driven language translation apps have become ubiquitous on smartphones and other devices. These apps leverage machine learning algorithms to analyze large datasets of language pairs, enabling them to accurately translate text and speech in real-time. Users can effortlessly translate documents, messages, or conversations, empowering them to engage with diverse language communities seamlessly.
- Real-Time Translation Services: AI has facilitated real-time translation services that enable multilingual communication during events, meetings, or international conferences. Real-time translation devices, equipped with sophisticated language models, can instantly convert spoken words from one language to another, facilitating smooth communication in diverse settings.
- Enhanced Accuracy and Contextual Understanding: AI language translation

algorithms continuously learn from vast amounts of linguistic data, improving their accuracy and contextual understanding over time. This adaptive learning process allows AI translation tools to interpret idiomatic expressions, cultural nuances, and evolving language trends more effectively, resulting in more natural and meaningful translations.

Benefits of AI-Powered Language Translation

- Facilitating Cross-Cultural Communication: AI-powered language translation fosters cross-cultural understanding by enabling people from different linguistic backgrounds to communicate effectively. This enhanced communication breaks down language barriers and promotes international collaboration, fostering global cooperation.
- Expanding Market Reach: For businesses, AI language translation offers the ability to reach a broader audience by breaking down language barriers in marketing, customer support, and international

expansion efforts. This opens up new markets and opportunities for growth.

- Enabling Access to Information: AI language translation ensures that individuals have access to information, education, and resources, regardless of the language in which they are originally available. This democratization of information fosters knowledge-sharing and empowers individuals globally.

AI-Powered Accessibility for Individuals with Disabilities

AI's potential extends beyond language translation to enhance accessibility for individuals with disabilities. By leveraging advanced machine learning techniques, AI can empower individuals with disabilities to engage more effectively with the world around them.

- Text-to-Speech (TTS) Capabilities: AI-driven TTS systems can convert written text into spoken words, assisting individuals with visual impairments, dyslexia, or other reading difficulties. These systems use natural language processing and neural networks to produce human-like speech, enhancing the accessibility of written content.

- Speech-to-Text (STT) Capabilities: For individuals with speech impairments or conditions that hinder verbal communication, AI-powered STT systems offer a lifeline. These systems convert spoken words into written text, allowing individuals to communicate through written means effectively.

- Smart Assistive Devices: AI technology has enabled the development of smart assistive devices that cater to the specific needs of individuals with disabilities. These devices can be controlled through voice commands or customized interfaces, providing greater independence and control over daily activities.

Benefits of AI-Powered Accessibility

- Promoting Inclusion: AI-powered accessibility tools empower individuals with disabilities to participate more actively in various aspects of life, fostering greater inclusivity and reducing barriers to education, employment, and social interaction.

- Enhancing Learning Opportunities: With AI-driven TTS and STT capabilities, educational materials and resources become accessible to students with diverse learning needs. This inclusivity ensures that all students can engage with the same information and educational content.

- Improving Workplace Productivity: AI-powered accessibility tools enable employees with disabilities to contribute effectively to the workplace, enhancing their productivity and enabling them to reach their full potential.

- Personalized Assistive Solutions: AI technology allows for the development of personalized assistive solutions, catering to the unique needs of individuals with disabilities. These tailored solutions can significantly improve their quality of life and independence.

Ethical Considerations and Challenges

Despite the transformative potential of AI in language translation and accessibility, several ethical considerations and challenges must be addressed:

- Data Privacy and Security: AI-powered systems often rely on large datasets for training. Ensuring the privacy and security of sensitive user data becomes crucial, especially when dealing with personal information or communication.
- Bias and Fairness: AI algorithms can inherit biases present in the training data, potentially perpetuating societal prejudices. Developers must take proactive steps to identify and mitigate such biases to ensure fairness and equity in AI applications.
- Accuracy and Reliability: While AI translation has made great strides, achieving complete accuracy and contextual understanding across all languages remains a challenge. The potential for miscommunication or misinterpretation may persist, necessitating ongoing improvements in AI language models.
- Digital Divide: The adoption of AI-powered tools is often contingent on access to technology and the internet. Addressing the digital divide ensures that individuals from all socioeconomic backgrounds can benefit from AI-driven accessibility solutions.

. . .

AI-powered language translation and accessibility tools have emerged as transformative forces, breaking down language barriers and enhancing inclusivity for individuals with disabilities. The ability to communicate effectively across languages fosters cross-cultural understanding and collaboration on a global scale. Furthermore, AI-driven accessibility solutions empower individuals with disabilities, offering them new opportunities for education, employment, and social engagement. However, developers and stakeholders must address ethical considerations, such as data privacy, bias, and reliability, to ensure the responsible and equitable implementation of AI technology. With continued advancements and ethical diligence, AI-driven language translation and accessibility promise to create a more connected and inclusive world for all.

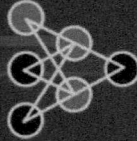

13

CASE STUDY 3
AI IN EDUCATION - PERSONALIZED LEARNING

PERSONALIZED LEARNING HAS EMERGED AS A critical approach to address educational inequities and cater to the diverse needs of students in today's classrooms. The traditional one-size-fits-all education model often fails to meet individual learning preferences and abilities, leading to disparities in academic achievement. Artificial intelligence (AI) has revolutionized the education landscape by enabling the creation of personalized learning experiences. This case study delves into the application of AI in personalized learning, exploring how AI algorithms analyze student data to customize instructional content and provide individualized feedback. By adapting to students' unique needs and abilities, AI-powered personalized learning platforms foster engagement, support academic growth, and promote educational equity. (Baker et al., 2008).

. . .

The Role of AI in Personalized Learning

- Analyzing Student Data: AI-powered personalized learning platforms rely on advanced data analysis to understand each student's strengths, weaknesses, and learning preferences. Student data may include academic performance, learning progress, assessments, and even non-cognitive factors such as motivation and engagement.

- Tailoring Instructional Content: Armed with comprehensive student data, AI algorithms customize instructional content to suit individual learning styles and abilities. This tailored approach ensures that students receive content at their appropriate skill level and pace, enhancing comprehension and retention.

- Providing Individualized Feedback: AI-driven systems deliver timely and targeted feedback to students, highlighting areas for improvement and offering personalized recommendations for their learning journey. This feedback loop fosters a growth mindset, motivating students to take ownership of their learning process.

Benefits of AI-Powered Personalized Learning

- Addressing Learning Gaps: Personalized learning using AI can identify and address learning gaps, allowing educators to intervene early and provide additional support to struggling students. This proactive approach reduces achievement disparities and prevents students from falling behind.
- Catering to Diverse Learning Styles: AI algorithms can adapt instructional materials to suit various learning styles, accommodating students' unique preferences and enhancing their overall learning experience.
- Promoting Student Engagement: Personalized learning captures students' interests and maintains their engagement by presenting content in a meaningful and relevant manner. Increased engagement leads to higher motivation and improved learning outcomes.
- Encouraging Self-Directed Learning: AI-powered personalized learning fosters self-directed learning habits, empowering students to take charge of their education

and pursue areas of interest independently.

AI-Powered Personalized Learning Platforms in Practice

- Adaptive Learning Systems: AI-driven adaptive learning platforms offer dynamic, responsive learning experiences that adjust in real-time based on individual student responses. These platforms continuously analyze performance and provide appropriate challenges, keeping students engaged and in their optimal learning zone.
- Intelligent Tutoring Systems: Intelligent tutoring systems utilize AI to provide personalized support and guidance, simulating the role of a human tutor. These systems can answer students' questions, explain concepts, and offer feedback, promoting deeper understanding and mastery of subjects.
- Learning Analytics: AI-driven learning analytics platforms gather data on student engagement, progress, and performance, helping educators identify patterns and trends to inform their instructional strategies. These insights enable teachers

to tailor their lessons to meet individual and group needs effectively.

Promoting Educational Equity through AI-Powered Personalized Learning

- Reducing Achievement Gaps: Personalized learning powered by AI can narrow achievement gaps by providing targeted interventions and support to students who may be at risk of falling behind. By addressing individual needs, these platforms ensure that all students receive the necessary resources to succeed academically.
- Catering to Diverse Student Populations: AI-driven personalized learning is especially beneficial for diverse student populations, including English language learners (ELLs), students with disabilities, and those from disadvantaged backgrounds. These platforms offer tailored learning experiences that accommodate diverse learning needs, ensuring equitable access to quality education.
- Enhancing Teacher Effectiveness: AI-powered personalized learning does not replace teachers but empowers them to be

more effective educators. By providing valuable insights and automating administrative tasks, AI enables teachers to focus on individualized instruction and better support their students.

Challenges and Ethical Considerations

While AI-powered personalized learning holds tremendous promise, it also poses challenges and ethical considerations that require careful consideration:

- Data Privacy and Security: AI algorithms rely on vast amounts of student data, raising concerns about data privacy and security. Educational institutions must prioritize safeguarding student information and ensure compliance with data protection regulations.
- Algorithmic Bias: AI algorithms may inadvertently perpetuate biases present in historical data, potentially leading to unequal treatment of certain student groups. Developers must continuously monitor and address algorithmic bias to ensure fair and equitable personalized learning experiences.
- Digital Divide: AI-powered personalized learning relies on access to technology and

the internet, potentially exacerbating existing disparities in access to education. Efforts must be made to bridge the digital divide and ensure all students have equitable access to personalized learning opportunities.

AI-powered personalized learning has emerged as a transformative tool to address educational inequities and meet the diverse needs of students. By analyzing student data, customizing instructional content, and providing individualized feedback, AI algorithms create dynamic and engaging learning experiences that foster academic growth and promote educational equity. Implementing AI-driven personalized learning platforms not only reduces achievement gaps but also empowers students to become self-directed learners, better preparing them for the challenges of the future. However, ethical considerations, such as data privacy, algorithmic bias, and the digital divide, must be carefully addressed to ensure responsible and equitable implementation. With continued advancements in AI technology and a commitment to ethical practices, personalized learning powered by AI holds the potential to revolutionize education and create more inclusive and equitable learning environments.

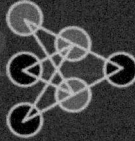

14

CASE STUDY 4

AI FOR HEALTHCARE DIAGNOSIS AND TREATMENT

TIMELY AND ACCURATE DIAGNOSIS IS AT THE CORE of effective healthcare delivery. Advancements in artificial intelligence (AI) have paved the way for innovative diagnostic tools and decision-support systems that can analyze medical images, patient records, and clinical data to aid in disease detection and treatment planning. This case study delves into the transformative impact of AI in the healthcare sector, particularly in improving diagnostic accuracy, reducing errors, and enabling personalized treatment approaches. By harnessing the power of AI, healthcare providers can enhance patient outcomes, reduce healthcare disparities, and ensure equitable access to quality care. (Obermeyer et al., 2019).

AI in Diagnostic Imaging

- Automated Image Analysis: AI-powered diagnostic tools can analyze medical images, such as X-rays, MRIs, and CT scans, with remarkable accuracy. By leveraging deep learning algorithms, these systems can identify abnormalities, tumors, and other conditions that may be missed by human radiologists.

- Early Disease Detection: AI algorithms can detect subtle signs of diseases at an early stage, allowing for timely intervention and improved patient outcomes. Early detection is particularly crucial in diseases like cancer, where early treatment significantly impacts survival rates.

- Image Segmentation and Quantification: AI can assist in image segmentation and quantification, aiding in precise measurements of tumor size, growth, or changes over time. This level of detail enhances treatment planning and monitoring.

Benefits of AI in Diagnostic Imaging

- Improved Diagnostic Accuracy: AI-based diagnostic tools complement the expertise of healthcare professionals, reducing the chances of misdiagnosis and improving overall diagnostic accuracy.

- Faster Diagnosis: AI algorithms can analyze medical images swiftly, expediting the diagnostic process and enabling healthcare providers to initiate treatment promptly.
- Enhanced Workflows: AI-based diagnostic systems streamline workflows by automating time-consuming tasks, allowing radiologists to focus on complex cases and patient care.
- Access to Expertise: AI can be particularly beneficial in areas with a shortage of specialized healthcare professionals, providing access to expert-level diagnostics in underserved regions.

AI in Clinical Decision Support

- Data Analysis and Predictive Modeling: AI can analyze vast amounts of patient data, including medical histories, lab results, and treatment outcomes, to identify patterns and trends. Predictive modeling helps in assessing patient risk factors and potential treatment responses.
- Personalized Treatment Plans: AI-based clinical decision support systems enable

personalized treatment plans by considering individual patient characteristics, medical history, and genetic information. This approach enhances treatment efficacy and reduces adverse effects.

- Drug Discovery and Development: AI accelerates drug discovery by analyzing complex biological data, identifying potential drug targets, and predicting drug efficacy more efficiently than traditional methods.

Benefits of AI in Clinical Decision Support

- Evidence-Based Treatment: AI-driven clinical decision support systems integrate the latest medical research and best practices, supporting healthcare providers in making evidence-based treatment decisions.
- Reduced Medical Errors: AI algorithms help mitigate medical errors, such as adverse drug reactions or incorrect treatment dosages, by providing real-time decision support based on accurate patient data.

- Patient-Centered Care: By tailoring treatment plans to individual patients, AI promotes patient-centered care, considering patients' preferences, values, and unique healthcare needs.

AI and Personalized Medicine

- Genomic Data Analysis: AI can analyze vast genomic datasets to identify genetic variations associated with specific diseases or treatment responses. This knowledge informs personalized treatment strategies and targeted therapies.
- Disease Risk Assessment: AI algorithms can assess an individual's risk of developing certain diseases based on genetic, lifestyle, and environmental factors. This information helps healthcare providers implement preventive measures and early intervention strategies.
- Predictive Modeling for Treatment Response: AI can predict how individual patients will respond to specific treatments, enabling physicians to select the most effective therapies for each patient.

Benefits of AI in Personalized Medicine

- Targeted Therapies: AI-driven personalized medicine enables the development of targeted therapies that address the unique characteristics of individual patients, resulting in higher treatment success rates.
- Reducing Treatment Side Effects: Personalized medicine ensures that patients receive treatments that are best suited for their genetic makeup and health conditions, reducing the likelihood of adverse side effects.
- Better Resource Allocation: By identifying patients who are more likely to benefit from specific treatments, AI-driven personalized medicine optimizes resource allocation and reduces healthcare costs.

Challenges and Ethical Considerations

- Data Privacy and Security: The widespread use of AI in healthcare generates massive amounts of patient data, raising concerns about data privacy and security. Healthcare providers must prioritize protecting patient information and comply with data protection regulations.

- Algorithmic Bias: AI algorithms can inherit biases from training data, potentially leading to disparities in healthcare delivery. Ensuring fairness and equity in AI-driven healthcare applications is paramount to avoid exacerbating existing healthcare disparities.
- Patient Acceptance and Trust: AI's role in healthcare requires patient acceptance and trust. Healthcare providers must transparently communicate the benefits and limitations of AI-driven solutions to build patient confidence in these technologies.

AI's integration into healthcare has revolutionized diagnostic practices, clinical decision-making, and personalized treatment approaches. By automating image analysis, enhancing diagnostic accuracy, and providing decision support based on vast patient data, AI augments healthcare professionals' capabilities, leading to better patient outcomes and reduced medical errors. In personalized medicine, AI-driven genomic analysis and predictive modeling enable targeted therapies and more effective treatment strategies. However, ethical considerations, such as data privacy, algorithmic bias, and patient acceptance, must be diligently addressed to ensure the responsible and equitable use of AI in healthcare. With continued advancements in AI technology and a commitment to ethical practices, the healthcare sector stands to benefit significantly from the transformative potential of AI in improving diagnostics and personalized treatment.

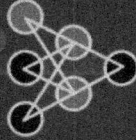

15

CASE STUDY 5

AI FOR SOCIAL MEDIA MODERATION

SOCIAL MEDIA PLATFORMS FACE SIGNIFICANT challenges in moderating content and preventing the spread of harmful or offensive material. In this case study, we examine how AI is employed to enhance social media moderation. AI algorithms can analyze user-generated content, detect potential violations, and flag inappropriate or harmful posts for review. This technology helps ensure safer online environments, protect users from harassment and hate speech, and promote inclusive and respectful online interactions (Fiesler et al., 2018).

In today's digital age, social media has become an integral part of modern communication, enabling people to connect, share information, and express their thoughts freely. However, with the vast amount of user-generated content being posted every second, social media platforms face significant challenges in maintaining a safe and inclusive environment. Harmful or offensive material, such as hate speech, harassment, and misinformation, can spread rapidly, posing serious threats to user well-being and platform credibility. To address these issues, many social media platforms have turned to artificial intelligence (AI) to enhance their moderation efforts. This case study explores how AI algorithms are employed to detect potential violations, flag inappropriate content, and ultimately foster safer and more respectful online inter-actions.

THE ROLE OF AI IN SOCIAL MEDIA MODERATION

AI-powered content moderation algorithms have emerged as a vital tool for social media platforms in the fight against harmful content. These algorithms leverage machine learning techniques to analyze vast amounts of user-generated content, identifying patterns and trends indicative of potential violations. Unlike human moderators who are limited by time and resources, AI algorithms can continuously and effi-ciently process vast amounts of content, ensuring that potentially harmful posts are addressed promptly.

Detecting Harmful Content

AI algorithms are trained on large datasets comprising examples of harmful or inappropriate content. Through a process known as supervised learn-ing, the algorithms learn to recognize patterns in the data and distinguish between acceptable and harmful content. For instance, they can detect hate speech, cyberbullying, or graphic violence. This ability to identify harmful content helps platforms take swift action to protect users from encountering distressing material.

Flagging Potential Violations

Once the AI algorithms detect potentially harmful content, they flag it for human review. This process helps ensure that false positives are minimized, as human moderators can contextualize content and make nuanced judgments that algorithms might miss. The combination of AI and human oversight allows for a more accurate and balanced approach to content moderation.

BENEFITS OF AI-DRIVEN SOCIAL MEDIA MODERATION

Enhancing User Safety

By using AI algorithms to detect and flag harmful content, social media platforms can create safer online environments for their users. Swift identification and removal of offensive material help prevent users from encountering harmful content that may trigger anxiety, fear, or emotional distress. Moreover, addressing harassment and hate speech can contribute to reducing online toxicity and promoting positive interactions.

Promoting Inclusivity and Respectful Interactions

AI-driven moderation not only focuses on removing harmful content but also on promoting inclusivity and respectful interactions. Algorithms can be programmed to identify and address discriminatory language, enabling platforms to foster an inclusive atmosphere where diverse voices are valued and protected from discrimination.

Scaling Moderation Efforts

The volume of user-generated content on social media platforms is staggering, making manual moderation nearly impossible to handle effectively. AI-powered content moderation scales effortlessly, processing millions of posts within milliseconds. This scalability ensures that the platform can keep up with the ever-increasing user base and content creation, safeguarding users from harmful content.

CHALLENGES AND ETHICAL CONSIDERATIONS

While AI-driven social media moderation offers significant benefits, it also poses several challenges and ethical considerations.

Algorithmic Bias

AI algorithms are only as effective as the data on which they are trained. If the training data is biased or unrepresentative, the algorithms may exhibit bias in their decision-making, potentially leading to the uneven treatment of certain user groups. For example, an algorithm might disproportionately flag content from marginalized communities or misclassify innocuous posts as harmful. Developers must be vigilant in addressing bias and regularly updating their models to improve fairness.

Content Nuances

AI algorithms often struggle with understanding the nuances of language and context. Sarcasm, satire, and cultural references may be misinterpreted, leading to erroneous content moderation decisions. Human moderation is essential for interpreting complex posts that algorithms may struggle to grasp fully.

Freedom of Expression

Striking a balance between moderating harmful content and upholding freedom of expression is a delicate task. Overzealous content moderation could lead to unnecessary censorship and limit the exchange of ideas, potentially stifling meaningful conversations. Platforms must develop transparent guidelines and provide clear explanations for content removal to maintain users' trust.

AI-driven social media moderation is an invaluable asset in tackling the challenges posed by harmful or offensive content on online platforms. By leveraging the efficiency of AI algorithms and the insights of human moderators, platforms can create safer spaces for users, foster inclusivity, and promote respectful interactions. However, developers must remain vigilant in addressing algorithmic bias, understanding content nuances, and upholding users' freedom of expression. As technology continues to advance, AI-driven social media moderation will play an increasingly crucial role in shaping the future of online discourse and user experiences.

16

CASE STUDY CLOSING REMARKS

THE CASE STUDIES HAVE PROVIDED A RANGE OF case studies that demonstrate the tangible impact of AI in promoting diversity, inclusion, and equity. From AI-powered hiring practices to personalized learning, healthcare diagnostics, social media moderation, and language translation, these examples highlight the transformative potential of responsible AI implementation. By leveraging AI in various domains, we can address societal challenges, reduce disparities, and create more inclusive and equitable environments. These case studies serve as inspiration for further exploration and innovation, illustrating the power of AI to positively impact our lives and shape a better future for all. In the final chapter, we will summarize the key takeaways from the case studies and draw over-arching insights that shed light on the potential of AI to drive social change and advance equity.

Throughout these case studies, a common thread emerges: responsible and ethical AI implementation is crucial for achieving positive outcomes. It is essential to recognize that AI is a tool that reflects the values and biases of its creators and the data it is trained on. Careful attention must be paid to data collection, algorithm design, and ongoing monitoring to mitigate bias and ensure fairness.

Additionally, collaboration and interdisciplinary approaches are key to leveraging AI for social equity. These case studies involve partnerships between technologists, domain experts, and affected communities. By involving diverse voices and perspectives, we can address the nuances and complexities of social inequities, avoiding the pitfalls of one-size-fits-all solutions.

As we reflect on the transformative potential of AI, it is important to strike a balance between embracing innovation and safeguarding against unintended consequences. Ethical considerations, transparency, and accountability must guide AI development and deployment. Regulatory frameworks and guidelines should be established to ensure that AI technologies are used responsibly and in ways that uphold social values and human rights.

The case studies presented in this chapter serve as beacons of hope, showcasing how AI can be harnessed as a force for positive change. They demonstrate that by intentionally and ethically applying AI, we can address systemic inequities, empower marginalized communities, and create a more inclusive and just society.

In the final chapter, we will synthesize the key insights from the entire book and offer recommendations for individuals, organizations, and policymakers to navigate the intersection of race, gender, neuroscience, unconscious bias, and artificial intelligence. Together, we can shape a future where AI becomes a powerful tool in the pursuit of equity, fairness, and social progress.

17

CONCLUSION

CHARTING A COURSE FORWARD

THE CONCLUSION SERVES AS THE CULMINATION OF our exploration into the intersection of race, gender, neuroscience, unconscious bias, and artificial intelligence. Throughout this book, we have examined the historical context, dissecting the underlying mechanisms of bias, and explored the transformative potential of AI in addressing social inequities. In this final chapter, we will summarize the key takeaways and offer recommendations for charting a course forward towards a more equitable and inclusive world shaped by AI.

One of the fundamental aspects of addressing bias and promoting equity is cultivating awareness. We must acknowledge and confront our own biases, both conscious and unconscious. By fostering self-reflection and engaging in open dialogue, we can gain insights into the ways in which our perceptions and actions are shaped by societal influences. Building awareness is the first step towards dismantling systemic biases and creating a more inclusive society.

Education plays a pivotal role in equipping individuals with the knowledge and skills to navigate the complexities of the modern world. To prepare Gen Z black girls and women for the challenges and opportunities brought by AI, it is crucial to prioritize education and training. This includes integrating AI literacy into curricula, promoting STEM education and career pathways, and fostering critical thinking skills to navigate the ethical implications of AI. Empowering individuals with the tools to understand, shape, and utilize AI can pave the way for increased representation and influence in AI development.

Diverse perspectives and experiences are essential for creating AI systems that are unbiased, fair, and inclusive. Encouraging diversity within AI teams, including racial and gender diversity, can help challenge assumptions, identify potential biases, and develop more comprehensive solutions. Organizations should actively work towards creating inclusive environments that attract and retain diverse talent. Furthermore, involving marginalized communities in the design and development of AI technologies ensures that their needs and perspectives are taken into account.

To ensure that AI is harnessed for positive social impact, it is crucial to establish ethical guidelines and responsible AI practices. Policymakers, industry leaders, and researchers should collaborate to develop and implement regulations that address bias, fairness, transparency, and accountability in AI systems. Ongoing monitoring and evaluation are necessary to assess the impact of AI technologies on different communities and to make necessary adjustments to mitigate harm and ensure equitable outcomes.

Addressing the complex challenges at the intersection of race, gender, neuroscience, unconscious bias, and AI requires collaborative efforts. Partnerships between academia, industry, policymakers, community organizations, and affected communities are key to developing comprehensive and contextually relevant solutions. By combining expertise and resources, we can foster innovation, exchange best practices, and collectively work towards creating an inclusive and equitable AI ecosystem.

In closing, this book has explored the intricate relationship between race, gender, neuroscience, unconscious bias, and artificial intelligence. We have examined the historical legacy of discrimination, revealing the mechanisms of bias, and highlighted the potential of AI to address social inequities. By leveraging awareness, education, diversity, ethics, and collaboration, we can chart a course forward towards a future where AI serves as a force for positive social change.

The responsibility lies with each of us, individually

and collectively, to actively contribute to the creation of an equitable and inclusive world. By understanding the power and limitations of AI, we can shape its development and deployment in ways that uplift marginalized communities, challenge biases, and promote justice. Let us embark on this journey together, driven by a shared commitment to social equity and guided by the transformative potential of AI.

Together, we can build a future where technology becomes a catalyst for a more just and inclusive society. It is through our collective actions and intentional efforts that we can ensure that AI technologies are harnessed to address the historical legacies of racism, sexism, and other forms of discrimination.

As we move forward, it is essential to remain vigilant and continuously evaluate the impact of AI systems on marginalized communities. Ongoing research and data collection should be conducted to assess whether AI interventions are achieving their intended goals or inadvertently perpetuating bias and inequity. This information can inform iterative improvements and course corrections to ensure that AI technologies are aligned with the principles of fairness, transparency, and accountability.

Furthermore, it is crucial to engage in ongoing dialogue and collaboration between stakeholders from diverse backgrounds. By actively seeking the perspectives of those who have historically been marginalized or underrepresented, we can develop AI systems that better reflect the needs and values of all members of society. This collaborative approach fosters a sense of ownership, empowerment, and shared responsibility for the development and deployment of AI technologies.

Ultimately, the journey towards an equitable and inclusive AI future requires a commitment to lifelong learning and adaptability. As technology evolves, new challenges and opportunities will arise. Staying informed about the latest advancements, engaging in critical discussions, and being open to innovation will enable us to navigate these changes with an equity lens.

In conclusion, the intersection of race, gender, neuroscience, unconscious bias, and artificial intelligence presents both challenges and opportunities. Through awareness, education, diversity, ethics, and collaboration, we can harness the power of AI to promote social justice, empower marginalized communities, and foster a more inclusive society. By charting a course forward based on these principles, we can shape a future where AI technologies contribute to the creation of a world that is fair, just, and equitable for all. Together, let us embark on this transformative journey and leave a positive impact for generations to come.

BIBLIOGRAPHY

Alken, Peter. "Alam R. et al., 2021: Contemporary Considerations in the Management and Treatment of Lower Pole Stones." STORZ MEDICAL - the Shock Wave Company, November 24, 2021. https://www.storzmedical.com/en/disciplines/literature-data base-blog/database-lithotripsy/alam-r-et-al-2021.

Amodio, David M. "The Neuroscience of Prejudice and Stereotyping." *Nature Reviews Neuroscience* 15, no. 10 (September 4, 2014): 670–82. https://doi.org/10.1038/nrn3800.

Bender, Emily M., Timnit Gebru, Angelina McMillan-Major, and Shmargaret Shmitchell. *On the Dangers of Stochastic Parrots*, 2021. https://doi.org/10.1145/3442188.3445922.

Blair, Clancy. "School Readiness: Integrating Cognition and Emotion in a Neurobiological Conceptualization of Children's Functioning at School Entry." *American Psychologist* 57, no. 2 (February 1, 2002): 111–27. https://doi.org/10.1037/0003-066x.57.2.111.

Buolamwini, Joy, and Timnit Gebru. "Gender Shades: Intersectional Accuracy Disparities in Commercial Gender Classification." *Machine Learning Research*, January 21, 2018, 77–91. http://proceedings.mlr.press/v81/buolamwini18a/buolamwini18a.pdf.

Crawford, Brett E., and M. Tina Dacin. "Policing Work: Emotions and Violence in Institutional Work." *Organization Studies* 42, no. 8 (July 16, 2020): 1219–40. https://doi.org/10.1177/0170840620941614.

Crenshaw, Kimberle. "Demarginalizing the Intersection of Race and Sex: A Black Feminist Critique of Antidiscrimination Doctrine, Feminist Theory and Antiracist Politics." Chicago Unbound, n.d. https://chicagounbound.uchicago.edu/uclf/vol1989/iss1/8.

Cvencek, Dario, Andrew N. Meltzoff, and Anthony G. Greenwald.

"Math-Gender Stereotypes in Elementary School Children." *Child Development* 82, no. 3 (March 9, 2011): 766–79. https://doi.org/10.1111/j.1467-8624.2010.01529.x.

Dasgupta, Nilanjana, and Anthony G. Greenwald. "On the Malleability of Automatic Attitudes: Combating Automatic Prejudice with Images of Admired and Disliked Individuals." *Journal of Personality and Social Psychology* 81, no. 5 (January 1, 2001): 800–814. https://doi.org/10.1037/0022-3514.81.5.800.

Devine, Patricia G. "Stereotypes and Prejudice: Their Automatic and Controlled Components." *Journal of Personality and Social Psychology* 56, no. 1 (January 1, 1989): 5–18. https://doi.org/10.1037/0022-3514.56.1.5.

"Education at a Glance 2019," n.d. https://www.oecd-ilibrary.org/education/education-at-a-glance-2019_f8d7880d-en.

European Commission. "European Commission, Official Website," n.d. https://commission.europa.eu/index_en.

Freeman, Jim, and Nalini Ambady. "A Dynamic Interactive Theory of Person Construal." *Psychological Review* 118, no. 2 (January 1, 2011): 247–79. https://doi.org/10.1037/a0022327.

Galinsky, Adam D., Erika V. Hall, and Amy J. C. Cuddy. "Gendered Races." *Psychological Science* 24, no. 4 (March 8, 2013): 498–506. https://doi.org/10.1177/0956797612457783.

"Gillespie, T. (2018). Custodians of the Internet Platforms, Content Moderation, and the Hidden Decisions That Shape Social Media. London Yale University Press. - References - Scientific Research Publishing," n.d. https://www.scirp.org/(S(czeh2tfqw2orz553k1wor45))/reference/referencespapers.aspx?referenceid=2935670.

Gillespie, Tarleton. "Content Moderation, AI, and the Question of Scale." *Big Data & Society* 7, no. 2 (July 1, 2020): 205395172094323. https://doi.org/10.1177/2053951720943234.

Greenwald, Anthony G., and Linda Hamilton Krieger. "Implicit Bias: Scientific Foundations." *California Law Review* 94, no. 4 (July 1, 2006): 945. https://doi.org/10.2307/20439056.

Hart, Allen J., Paul J. Whalen, Lisa M. Shin, Sean C. McInerney, Håkan Fischer, and Scott L. Rauch. "Differential Response in the Human Amygdala to Racial Outgroup vs Ingroup Face Stimuli." *Neuroreport* 11, no. 11 (August 1, 2000): 2351–54. https://doi.org/10.1097/00001756-200008030-00004.

Influence Watch. "World Economic Forum - InfluenceWatch." InfluenceWatch, April 19, 2022. https://www.influencewatch.org/non-profit/world-economic-forum/.

Jobin, Anna, and Effy Vayena. "The Global Landscape of AI Ethics Guidelines." *Nature Machine Intelligence* 1, no. 9 (September 2, 2019): 389–99. https://doi.org/10.1038/s42256-019-0088-2.

Lai, Calvin K., Maddalena Marini, Steven A. Lehr, Carlo Cerruti, Jiyun Elizabeth L. Shin, Jennifer A. Joy-Gaba, Arnold K. Ho, et al. "Reducing Implicit Racial Preferences: I. A Comparative Investigation of 17 Interventions." *N/A*, August 15, 2016. https://doi.org/10.31234/osf.io/ktgwv.

Lai, Calvin K., Allison L. Skinner, Erin Cooley, Sohad Murrar, Michael Brauer, Thierry Devos, Jimmy Calanchini, et al. "Reducing Implicit Racial Preferences: II. Intervention Effectiveness across Time." *Journal of Experimental Psychology: General* 145, no. 8 (August 1, 2016): 1001–16. https://doi.org/10.1037/xge0000179.

Madeira, Fábio, Youngsoo Park, Joon Seong Lee, Nicola Buso, Tamer Gur, Nandana Madhusoodanan, Prasad Basutkar, et al. "The EMBL-EBI Search and Sequence Analysis Tools APIs in 2019." *Nucleic Acids Research* 47, no. W1 (April 12, 2019): W636–41. https://doi.org/10.1093/nar/gkz268.

Shutts, Kristin, Elizabeth Brey, Leah A. Dornbusch, Nina Slywotzky, and Kristina R. Olson. "Children Use Wealth Cues to Evaluate Others." *PLOS ONE* 11, no. 3 (March 2, 2016): e0149360. https://doi.org/10.1371/journal.pone.0149360.

Slussareff, Michaela. "O'Neil, Cathy. 2016. *Weapons of Math Destruction: How Big Data Increases Inequality and Threatens Democracy* . Crown." *Cyber Orient* 16, no. 1 (June 1, 2022): 72–75. https://doi.org/10.1002/cyo2.26.

Steele, C. M., and Joshua Aronson. "Stereotype Threat and the Intellectual Test Performance of African Americans." *Journal of Personality and Social Psychology* 69, no. 5 (January 1, 1995): 797–811. https://doi.org/10.1037/0022-3514.69.5.797.

Wachter, Sandra, Brent Mittelstadt, and Luciano Floridi. "Why a Right to Explanation of Automated Decision-Making Does Not Exist in the General Data Protection Regulation." *Social Science Research Network*, January 1, 2016. https://doi.org/10.2139/ssrn.2903469.

Wing, Jeannette M. "Computational Thinking." *Communications of the ACM* 49, no. 3 (March 1, 2006): 33–35. https://doi.org/10.1145/1118178.1118215.

Yassin, Cherouk Amr Abdel Hakim. "Understanding Consumer Digital Consumption Behaviour in the Edge of Social Media Platforms." *Open Journal of Social Sciences* 09, no. 10 (January 1, 2021): 394–416. https://doi.org/10.4236/jss.2021.910028.

Zheng, Jingxu, Qing Zhao, Tian Tang, Jiefu Yin, Calvin D. Quilty, Genesis D. Renderos, Xiaotun Liu, et al. "Reversible Epitaxial Electrodeposition of Metals in Battery Anodes." *Science* 366, no. 6465 (November 1, 2019): 645–48. https://doi.org/10.1126/science.aax6873.

Dr. Misty D. Freeman grew up and lives in Alabama. She is an educator with over 20 years of experience and training in counseling, education, human resource management, and instructional technology. She is dedicated to educating others on unconscious bias, especially toward black girls and women of Gen Z. Due to the intersection of race and gender, black girls and women are more likely to experience unconscious bias. Her experience as a social caseworker, special education teacher, school administrator, and director of special education has provided her with opportunities many do not experience. She is keenly aware of the lack of unconscious bias understanding. She has committed to helping others define, identify, and disrupt unconscious bias. Also, Dr. Freeman is a thought leader, unconscious bias coach, and AI futurist. Her mission is to help others understand and transform their perspective of unconscious bias against girls and women of color at the intersection of learning, the workplace, and technology.

facebook.com/drmistydfreeman

instagram.com/drmistydfreeman

amazon.com/stores/author/B0C9VY77QN

linkedin.com/in/drmistydfreeman

www.ingramcontent.com/pod-product-compliance
Lightning Source LLC
Chambersburg PA
CBHW060249030426
42335CB00014B/1639